Two Models of Information

A Dissertation Presented to Cambridge State
University for the Requirements for the
Doctoral Degree in Electrical Engineering

June 14, 2001

By

Bradley Scott Tice

ISBN: 1-4033-4394-2 (e-book)
ISBN: 1-4033-4395-0 (Paperback)

This book is printed on acid free paper.

1stBooks – rev. 01/23/03

In Memory of

Willard Van Orman Quine*
(June 25, 1908 - December 25, 2000)

and

Claude Elwood Shannon**
(April 30, 1916 - February 24, 2001)

---

* Obituaries section of the <u>San Jose Mercury Newspaper</u>, section 5B, Friday, December 29, 2000 and Obituary section of <u>The Economist</u>, January 13, 2001, page 86.
** Short biography on Shannon is listed at http://www-groups.dcs.st-and.ac.uk/~history/Mathematicians/Shannon.html. Of interest is the statement by Robert Gallager in the article "Shannon Day at Bell Labs" that notes that Shannon was driven by 'curiosity' as the main criteria in selecting research problems and that this kind of research is now and 'endangered species' (Telater, 1998: 25). Another telling aside is that Gallager feels that if Shannon were starting his career today he would probably not get tenure at a major research university or a position at a major research lab (Telater, 1998: 25-26). This may, in part, be the reason why Lucent Technologies; i.e. Bell Labs, was almost sold to a French company in 2001. Lack of talented people produce few ground-breaking results and this, in turn, makes for a 'limited' future for the company. Telater, E. (1998) "Shannon Day at Bell Labs", <u>IEEE Information Theory Society Newsletter</u>, Summer 1998, Special Golden Jubilee Issue. On a similar note Michael Riordan and Lillian Hoddeson state in their well written account of the birth of the transistor, another Bell Labs invention, that by the 1960's Bell Labs slowly began to lose its' innovative edge (Riordan and Hoddeson, 1997:282). Riordan, M. and Hoddeson, L. (1997) <u>Crystal Fire: The Birth of the Information Age</u> [New York: W.W. Norton & Company].

Abstract

Shannon's Information Theory is explained and addressed in context to language. Algorithmic Information Theory is addressed in relation to randomness and complexity, with the work of Chaitin, Kolmogorov, and Solomonoff being reviewed with questions raised that deal with the fundamental levels of both randomness and complexity.

Preface

The first part of this dissertation deals with Claude Shannon's information theory. The second part of the dissertation is on aspects of randomness and complexity. They represent the mathematical fundamentals of information systems.

Acknowledgments

I would like to thank the Pacific Division of the American Association for the Advancement of Science [AAAS] for allowing me to present my ideas at their annual conferences over the years.

"By the year 2000 information theory may exist only in a few unread definitive treatises, preserved by college librarians as forlorn monuments to misspent lives"*

* Passage taken from page 320 of "Information theory after 18 years" by E.N. Gilbert, **Science**, April 15, 1966, pages 320-326.

Contents

Introduction

The first part of the dissertation deals with information theory. Shannon's Information Theory is explained and addressed in context to language. The second part of the dissertation is on aspects of randomness and complexity. In addressing randomness and complexity, the work of Chaitin, Kolmogorov, and Solomonoff is reviewed with questions raised that deal with the fundamental levels of both randomness and complexity.

Information Theory

Information theory, or more accurately statistical communication theory, is a communication model that has a signal chosen from a specific class that is to be transmitted through a channel, but the output of the channel is not determined by the input. The channel is described statistically by giving a probability distribution over the set of all possible outputs for each permissible input.

The output of the channel a signal is received and the identification of it as closely as possible to some property of the input signal (Ash, 1965: v). Ash has defined information theory as 'a communication system which has traditionally been represented by the block diagram' (Ash, 1965: 1). A communication system model represented by a series of block

diagrams is seen in Figure 1 (Ash, 1965: 1)*.

---

* I have designed a scheme for language recording and evaluation using the feedback system found in Shannon's communication system model that I have termed the 'Language Learning Loop' and I was honored to present it at the 50th Anniversary of Information Theory Conference at MIT in 1998.

"Language Learning Loop: A Feedback System" Paper presented in the recent results session of the 1998 IEEE International Symposium on Information Theory, ISIT, August 16-21, 1998 at MIT, Boston, MA. USA.

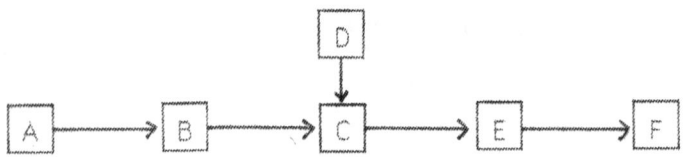

Block Diagram Codes

A=Source of Messages

B=Encoder

C=Channel

D=Noise

E=Decoder

F=Destination

Signal flow diagrams can also be used to depict systems, and feedback within these systems, and denote a linear hierarchy within a system as well (Tice, 1998).

Abramson states:

> The origins of information theory date back to Claude E., Shannon's publication of a paper in the Bell System Technical Journal in 1948. In terms of the colloquial meaning of information, Shannon's paper deals with the carriers of information-symbols- and not with information itself. It deals with communication and the means of communication rather than that elusive end product of communication- information (Abramson, 1963: 1-2).

The measure of self information is the binary unit of information associated with the 'bit'. The term 'bit' is a contraction of 'binary digit' as is in common usage today. The terms 'nat', or 'natural unit', and 'Hartley', in honor of R.V. Hartley as pioneer in communication theory, are not in general use today (Fano, 1961: 27). A bit is a self information of a binary digit when the two possible digits occur with equal probability. Hence the name of the unit of a binary digit or 'bit' (Fano, 1961: 36 and Gilbert, 1966: 321).

Of interest in information theory is the concept of 'entropy'. The self information of a message can be interpreted as the amount of information that must be provided about the message in order to specify it uniquely (Fano, 1965: 42). Hence the amount of information that must be provided on the average, in order to specify any particular message of an ensemble X, is the average value of the self information (Fano, 1961: 42). The self information of a particular symbol represents the amount of information that the symbol can provide about the message input to the decoder. The entropy of an ensemble of symbols represents the amount of information that a symbol can provide, on average, and, as such, it is a measure of the effectiveness with which the different symbols can be employed (Fano, 1961: 43).

Shannon defines both entropy and redundancy for his information theory model:

The ratio of the entropy of a source to the maximum value it could have while still restricted to the same symbols will be called its relative entropy. This, as will appear later, is the maximum compression possible when we encode into the same alphabet. One minus the relative entropy is the redundancy. (Shannon and Weaver, 1949: 56).

Shannon also defines continuous and discrete entropies:

There is one important difference between the continuous and discrete entropies. In the discrete case the entropy measures in an absolute way the randomness of the chance variable. In the continuous case the measurement is relative to the coordinate system. (Shannon and Weaver, 1949: 90).

Kahn notes from his informative book <u>The Code Breakers</u> (1967/1996):

Determining the percentage of redundancy begins with the calculation of a quantity called "entropy". Shannon borrowed this term from physics because the form of the equation that he developed for the amount of information in a set of utterances is identical with that used in physics to represent entropy. In both fields, entropy measures disorder, randomness, lack of structure. The greater the entropy the greater the chaos" (Kahn, 1967/1996: 759).

The fundamental measure of information can be defined as the change of probability of one of the events as a result of the observation of the other. In other words the logarithm of the ratio of the a posteriori probability and the a priori probability of the first event (Fano, 1961: 59). Such a measure is symmetrical in the two events, and that it can be regarded as a measure of the mutual information between the two events (Fano, 1961: 59).

Information theory demonstrates that messages from an information source that is not completely random can be compressed (Chaitin, 1970: 7). The definition of randomness is merely the converse of this fundamental theorem of information theory in that if in the absence of randomness in a message allows it to be coded into a shorter sequence, then the random messages must be those that cannot be coded into shorter messages (Chaitin, 1970: 7).

It is important to point out that the concept of redundancy is usually associated in a 'pejorative' sense in both the rhetorical and information theory definitions of the term (Birdwhistell, 1970: 85). Kahn (1967/1996) defines redundancy, as used in information theory, as being more symbols are being transmitted in a message than are actually needed to bear the information (Kahn, 1967/1996: 744). From the standpoint of this paper the term redundant or redundancy is to be associated with a feature of duplication only and not as a contextualized semantic element of social or behavioral traits.

It is interesting to note that the linguist Roman Jakobson seized the connection between information theory and his view of phonology and that it could be used to provide for a rigorous scientific basis for the interpretation and analysis of phonological systems (Anderson, 1985: 135 and Cherry, Halle, and Jakobson, 1953: 34-46). A singular problem with this is that while information theory is based on the optimization of code as an information channel such a goal for human or natural language is not a clear case (Anderson, 1985: 136). In fact, natural language is heavily oriented towards redundancy, as a core feature of the nature of human language (Anderson, 1985: 136).

Formal and Natural Languages

A Formal language is composed of signs and symbols that represent a mathematical representation of a language or a code.

These are usually 'computer' languages designed to be input, a question, into a computer and calculated in a series of steps or in parallel to produce an output, an answer to the question.

The brain is a enormously more complicated than even the most powerful computer. Most computers process information in a linear fashion. Discrete steps are taken, i.e. a series of sequential steps, to break down the process of calculation of a problem. Parallel processing computers deal with data in a simultaneous fashion that links several processing units together to compress the time a linear sequential computer would take to do the same task.

Natural Languages and Information Theory

Natural languages are redundant, i.e. they repeat common features with in the sound system corpus, and that this is an ideal trait in languages, but a problem in all other coded message systems. In 1948 Claude Shannon, a mathematician at AT&T's Bell Labs, published <u>The Mathematical Theory of Communication</u> that initiated the **Information Theory** revolution. Shannon was interested in the problem of maximizing the amount of information that you can transmit over a perfect communication channel such as a noisy telephone line. For any source of 'information' and 'communication channel', Shannon wanted to be able to determine theoretical maxima for

1.) data compression (Entropy [H])

2.) the transmission rate (Channel Capacity [C]).

Entropy measures the amount of information in a random variable. It is normally measured in bits. Entropy, the amount of information in a random variable, can be thought of as the average length of the message needed to transmit an outcome of that variable (Manning and Schutze, 1999: 61).

The Goal of communicating is to optimize the accuracy of the communicated message in the presence of 'noise' in the channel. **Compression** is the removal of all redundancy. **Redundancy** is the amount of repetition in the signal's message. A balance of compression and redundancy is the ideal of optimization in that the goal is to encode the message in such a way that it occupies minimal space while still containing enough redundancy to be able to detect and correct errors.

The central concept that characterizes a **channel** in information theory is its capacity. The channel capacity describes the rate at which one can transmit information through the channel with an arbitrarily low probability of being unable to recover the input from the output. **Mutual Information** is a symmetric, non-negative measure of the common information in two variables, in other words, the reduction of uncertainty of one random variable due to knowing about another variable.

Linguistics uses the **'noisy channel model'** that decodes the output to give the most likely input. Entropy is the measure of our uncertainty. The more we know about something, the the lower the entropy will be because we are less surprised by the outcome of the trial. If a model of a language captures more of the structure of a language, then the entropy of the model should be lower. Entropy is used as a measure of the quality of the language models used.

The Entropy of English

When the statistical effects extended over not more than eight letters the redundancy is about 50 per cent in such factors as

1.) the letter E has a high frequency of appearing.

2.) the letter H to follow the letter T

3.) the letter U to follow the letter Q

In ordinary literary English, up to 100 letters, the redundancy is about 75 per cent (Shannon, 1950/1993: 194). The most frequent word in English is [the] and the second most frequent word is [of].

Brown and Pietra (1992) have estimated the upper bound of 1.75 bits for the entropy of characters in printed English by using a construction of a word trigram model and then computing the cross-entropy between this model and a balanced sample of English text (Brown and Pietra, 1992: 31).

Information Theory and Phonology

It was thought in the 1950's and 1960's that Information Theory would provide a general rigorous scientific basis for the interpretation and analysis of phonological systems. Information Theory was used to present precise probabilities of the occurrence of particular segments, features, and sequences of features in a given corpus of linguistics text. The problem with this is that the way language is stored mentally, produced, and understood gives us little reason to believe that the principle of optimization and avoidance of redundancy as the fundamental role of human speech production. On the contrary, everything about language use seems to be characterized by massive amounts of redundancy as being essentially co-present with the supposedly more fundamental 'distinctive' elements of structure (Anderson, 1985: 134-139).

A Case for Information Theory in Phonology

I have designed a formal learning system using a system of recording devices that first initiates a vocal response from a L2, second language, student with the model or target L2 sound and then records that L2 response with the original target L2 sound.

This was termed a **Language Learning Loop** because it was based on a closed circuit of sound stimuli and sound response in L2 students learning a new sound system (Tice, 1997: 12). This systematic learning system was designed using Shannon's Information Theory of Communication and B.F. Skinner's theory of Stimuli-Response [S-R] learning behavior. This is one of the few appropriate uses of information theory in the area of phonology.

Algorithmic Information Theory

Algorithmic Information Theory [AIT] was invented by A.N. Kolmogorov and developed by R.J. Solomonoff and G. Chaitin. While Solomonoff defines that program-size complexity quantifies Occam's Razor by providing a numerical measure of the degree of simplicity of a scientific theory, it was Kolmogorov and Chaitin that, independently, came up with program-size complexity (Chaitin, 1999: 86). Program-size complexity can be stated that if a random string is incompressible, its program-size complexity is as large as possible for bit strings having that length (Chaitin, 1999: 86).

As stated in Beltrami (1999) "a string S is said to be algorithmically random if its algorithmic complexity is maximal, compared to the length of the string, meaning that it cannot be compressed by employing a more concise description other than writing out S in its entirety" (Beltrami, 1999: 94). The importance of Chaitin's work is that his theory of complexity-based definition of randomness works for both finite and infinite strings and is equivalent to Martin-Lof's (1966) theory for infinite strings (Chaitin, 1999: 87, Chaitin, 1975a: 337, Chaitin, 1966: 568, and Martin-Lof, 1966: 612-619).

Kolmogorov's paper (1965) describes the algorithmic approach as "the quantity of information "conveyed by an object" (x) "about an object" (y). It is not an accident that in the probabilistic approach this has led to a generalization to the case of continuous variables, for which the entropy is infinite, but in large number of cases, is finite. The real objects that we study are very (infinitely) complex, but the relationships between two separate objects diminish as the schemes used to describe them becomes simpler" (Kolmogorov, 1965: 4). From Bennett's (1998) on Kolmogorov's work:

Kolmogorov defined what he had meant by a "simple" law or formula and published a new quantitative measure of "information". In other words, rather than requiring the randomness of a sequence to be judged by absolute unpredictability, Kolmogorov would require only unpredictability by a small set of simple rules (Bennett, 1998: 163).

The ultimate random sequence in terms of this complexity measure would be one that could be

described only by naming the sequence itself,
element by element; no formula that is shorter than
the length of the sequence could be developed
(Bennett, 1998: 164).

Some salient points are made in Siegfried's book <u>The
Bit and the Pendulum</u> (2000) as he states:

Information theory as written down by Shannon is
not a theory of information content. Its a
quantitative theory of information" (Siegfried,
2000: 171).

Siegfried also notes:

The distribution between information quantity and
content has to do with the ability to compress the

description of a complex system. If a long string of bits can be compressed into a short computer program, there is a lot of regularity, but not very much information. From that point of view, high regularity means low complexity, and low information. But what about a message that can't be compressed? Such a message would be a random string of bits, for example, with no regularity. That message would take up the most space on your harddrive, so it would have lots of information, in Shannon's sense. But if it's a string of random digits, it wouldn't have much in the way of information content. Pure randomness produces a high information quantity for Shannon, and pure randomness also dominates measurements of algorithmic complexity."

(Siegfried, 2000: 168-169).

Some interesting questions arise when reviewing Gregory J. Chaitin's paper "Randomness and Mathematical Proof" (1975b).

> "Instructions given the computer must be complete and explicit, and they must enable it to proceed step by step without requiring that it comprehend the result of any part of the operation it performs. Such a program of instructions is an algorithm. It can demand any finite number of mechanical manipulations of numbers, but it cannot ask for judgments about their meaning" (Chaitin, 1975b: 47-48).

It is interesting to note that, in a way, all 'mechanical manipulations' of numbers ascertain both a quantitative; the obvious result of a calculation, but also qualitative; i.e. to evaluate a statistical quantity in a qualitative fashion, and hence for a rational opinion or judgment on that number.

"…this "incompressibility" is a property of all random numbers; indeed, we can proceed directly to define randomness in terms of incompressibility: A series of numbers is random if the smallest algorithm capable of specifying it to a computer has about the same number of bits of information as the series itself" (Chaitin, 1975b: 48).

Example (Chaitin, 1975b: 48):

Type	Numbers	Reduced
[Predictable]	01010101010101010101	[10 ten times]
[Random]	01101100110111100010	[same amount]

What if we grouped the numbers/bits into a series of 5 groups containing 4 characters? For ease of discussion the predictable group of numbers will be labelled A.) and the random group numbers B.).

A.) [0101] [0101] [0101] [0101] [0101]

B.) [0110] [1100] [1101] [1110] [0010]

We can reduce each group of numbers/bits into a specific type of grouping by using an alpha code to represent each type of frequency of numeration.

A.) a=0 followed by a 1, or ax5=01x10, or A.).

B.) a=0110

    b=1100

    c=1101

d=1110

e=0010

or a+b+c+d+e=B.).

Reducing both A.) and B.) to a two character series into an alpha code would produce:

A.) a=0 followed by a 1, or ax10.

B.) a=01

　　b=10

　　c=11

　　d=00

　　or [a+b] [c+d] [c+a] [c+b] [d+e]

　　or a+b+c(d+a+h) +d+e

Bradley S. Tice

Notice that although A.) stays the same, i.e. A=ax10, B is reduced from a 20 number/bit series into a group of 5 quad numbered groupings of numbers/bits that is further reduced to a binary code of characters using 4 alpha symbols to represent the four types of variations of the binary bits, i.e. 00, 11, 01, and 10, that when expressed mathematically combines the three internal groups, all designated by the alpha variable [c] in the head position, that reduces it into an equation of 3 groups from the original 'random' 20 numbers/bits series.

Of note is that if we reduce the 20 characters random series [B] into a 4 group alpha series of binary code states, i.e. [00], [11], [01], and [10],the level of randomness or 'uncertainty' is lowered as [c], number

1 followed by number 1 [11], repeats in the primary or head position in the middle three groupings in [B].

Knuth (1998) makes the observation that a truly random sequence will exhibit local nonrandomness (Knuth, 1998: 152). So can be said of the sub-dividing of a random string of numbers into subgroups of uniform bits that gives the impression of nonrandomness. This maybe stated as a 'natural' consequence of reduction to a 'uniform' or patterned sequence of a string of units or bits. We can formally propose such a process as:

A string of numbers that is random in nature will, upon sub-dividing into multiple equal units of measure of the whole, will produce a pattern of bits, that upon inspection, will have a uniform pattern to each sub-group of the divided whole.

Kolmogorov makes a similar observation in his paper (1969) when he states that "It is understood that the description of segments of an infinite sequence of such a nature can be significantly simplified in comparison with the standard description"

(Kolmogorov, 1969: 2).

Perceptions of 'Patternless' Numbers

Taking the binary bit strings:

A.) 01010101010101010101

B.) 01101100110111100010

From Chaitin:

The first is obviously constructed according to a single rule; it consists of the number 01 repeated ten times. If one were asked to speculate on how the series might continue, one could predict with considerable confidence that the next two digits would be 0 and 1. Inspection of the second series of digits yields no such comprehensive pattern. There is no obvious rule governing the formation of the number, and there is no rational way to guess the succeeding digits. The arrangement seems haphazard. In other words, the sequence appears to be a random assortment of 0's and 1's (Chaintin, 1975b, 47).

Why would we be able to 'predict' the next two numbers from (A), being 0 followed by a 1, and not predict with equal certitude, the 'promise' of being correct in one's prognostication based on previous patterns of information, when we are given an equally valuable amount of information in that, perhaps; a word best suited to predictions, (B) repeats itself! If (A) can repeat itself on a binary string level, hence 0 is followed by a 1, in a sequential order, for a total of ten repetitions, then why wouldn't (B) repeat its bit number count, in the same order and the same number, as (A), with the same degree of proposed accuracy? The question is more of perception rather than of statistical odds.

Non-Randomness then is a level of perceptual symmetry, i.e. duplication is a property of the frequency of regularity, the more regular the pattern, i.e. 01010101, the more it will continue as such, and randomness is the reverse, the greater divergence in regularity of a pattern will produce a tendency to 'create' new patterns. Non-Randomness equals a tendency to stay the same (the old pattern duplicated). Randomness equals the tendency to 'create' new patterns (no fixed sequence of numbers).

The problem seems to stem from 'what' a regular pattern is and 'how' that pattern will duplicate over an irregular pattern. If we take two examples from the paper, (A) is a binary sequence, or string, that repeats every other bit, i.e. 0 followed by 1 followed by the same sequence for a 20 character length total. (B) has an irregular pattern, i.e. not a strict duplication of the same binary code sequence, i.e. 01010101, but at the

Bradley S. Tice

20 bit/number level, it has a regular pattern, or at least a distinct pattern, just like (A).

And what is that regular 20 character length pattern?

01101100110111100010. Just like (A) has a 20 character length pattern, 01010101010101010101, (B) has a regular features that are not random at all, just not sequentially similar or as frequently as ordered.

Indeed, (B) is not 'patternless' at all. (B) has a strict rule formed pattern of binary bit code that breaks down as such:

0110 1100 1101 1110 0010

or

[0110] [1100] [1110] [0010]

or

[01]   [10]   [11]   [00]   [11]   [01]   [11]

[10]   [00]   [10]

If we take these binary bits two sequential places, to the right, i.e the first and second bit numbers starting at the left; 0 and 1, as a group and then followed by the remaining 9 groups of two bits, we have the following 4 binary bit code variations:

[01]

[10]

[11]

[00]

Rather than a random 'patternless' 20 bit number length, we have, in fact, an series or string of bit numbers that have a specific order and pattern that can be broken down into a series of 4 binary bit code

Bradley S. Tice

variations, that has a 20 character number limit, i.e. total per string, that is known, i.e. is not predicted but is a known or tangible property. Randomness is then in 'how' the pattern of a sequence of numbers is perceived, rather than a trait of the numbers themselves. There is randomness in numbers, but alot of that randomness maybe perceptual in nature. We may be placing more on 'form' than on 'function' regarding patterns in strings of numbers.

Alonzo Church, in a paper (1940), defines a random sequence:

An infinite sequence a1, a2,...of 0's and 1's is a random sequence if the two following conditions are satisfied:

(1) If f(r) is the number of 1's among the first r terms of a1, a2,...then f(r)/r approaches a limit p as r approaches infinity.

(2) If o is any effectively calculable function of positive integers, if b1=1, bn+1=2bn+an, cn=o(bn), and the integers n such that cn=1 form in order of magnitude an infinite sequence n1, n2,..., and if g(r) is the number of 1's among the first r terms of an1, an2,..., then g(r)/r approaches the same limit p as r approaches infinity (Church, 1940: 134).

While statistical probability is the domain of physical and natural sciences great work has been done in the field of economics. Early work by von Mises and later the work of von Neumann and Morgenstern, establishing the new field of game theory in their seminal work <u>Theory of Games and Economic Behavior</u> (1944), have set standards for this study that are current today. Bennett (1998) describes Von Mises's definition of randomness:

> Von Mises defined randomness in a sequence of observations in terms of the inability to devise a system to predict where in a sequence a particular observation will occur without prior knowledge of the sequence (Bennett, 1998: 161).

Randomness then, to Von Mises' conception, is based on the lack of knowledge of the future of a particular sequence in time of an observation. Randomness then deals with the ability to predict the future to some degree of accuracy of a predicted outcome.

Solomonoff Model

From Chaitin [regarding the Solomonoff Model]:

The model demands that the smallest algorithm, the one consisting of the fewest bits, be selected (Chaitin, 1975b: 49)

Chaitin continues with the Solomonoff Model:

Thus in the Solomonoff Model a theory that enables one to understand a series of observations is seen as a small computer program that reproduces the observations and makes predictions about possible future observations. The smaller the program, the more comprehensive the theory and

41

the greater the degree of understanding. Observations that are random cannot be reproduced by a small program and therefore cannot be explained by a theory (Chaitin, 1975b: 49).

The following is a 'Inductive Reasoning' example of a Solomonoff Model (Chaitin, 1975b: 49):

Observations   Predictions_____

                01010101010101010101

0101010101

                01010101010000000000

Theory	Size of Theory
Ten repetitions of 01	23 characters

Five repetitions of 01 followed by Ten 0's

What if we reduce the 'predictive' section to a more efficient binary bit code by using a space, i.e.. neither a 0 or a 1, as a signal for the computer to multiply the last number before the space, by the number that follows that space, hence a multiplication program that is formalized by three rules:

1.) The number before the space is the number to be multiplied.

2.) The number following the space is the multiplier.

3.) Two spaces concludes further multiplication procedures and hence returns the computer to the next operation.

Example: From the 'Prediction section of the Solomonoff Model:

Predictions_____

01010101010101010101

01 101 0 1010

Note: Five is the binary representation [101] and ten is the binary representation [1010] on a four character length.

Compare the original 'Predictions' section with the same section using the 'Space Multiplier'.

Original:

Predictions	Number of Spaces
01010101010101010101	20
01010101010000000000	20

Space Multiplier:

Predictions	Number of Spaces
01010101010101010101	20
01 101 0 1010	14

By doing this, we have reduced the amount of redundancy in the information code, reducing the level of 'randomness' in the signal, while maintaining the same amount of information, i.e. no entropy or degradation to the information content, and while not as simple as ten repetitions of 01, the level of randomness is reduced to a workable algorithm.

From this it can be shown that using the very same examples as used for Solomonoff's Model, 'randomness' was reduced by a 'programming' modification by the introduction of the 'space multiplier' that reduced the level of redundancy and the level of randomness because the 'actual' level of randomness was just a perceived level of complexity, i.e. that the level of complexity was doubled in the second string of binary bits, that required two different algorithm's rather than just one, but was still a reproducible and functional algorithm.

Again, it is the 'perception' of randomness rather than an actual model of such behavior that is the key to this functioning and that a small algorithm is not the answer. What is needed is an accurate algorithm, not one based on brevity of length.

If we take each of the number strings from the Solomonoff Model, an using only the first two binary bits in each string, the following would result:

A.) 01

B.) 01

If we used an algorithm based on these two binary bits, the following 'predictive' strings would result:

A.) 01010101010101010101

B.) 01010101010101010101

String [A] is correct.

String [B] is incorrect.

The smallest algorithm did not predict correctly string [B].

Again if we used the 'space multiplication' program we have for string [B]:

B.) 01010101010000000000

Which results in:

B.) 01 101 0 1010

We get:

B.) 01010101010000000000

In other words, it is the quality of information in an algorithm that is important, not the brevity of it, as such information is both quantitative, i.e. necessitates a certain amount of information to function, as well as qualitative, i.e. must accurately represent the information being imparted.

Using a variation of the three models from Solomonoff's paper "Formal Theory of Inductive Inference" (1964) some questions are raised about these inductive models.

List of Models:

1.) General Intuitive Model: Using Occasm's razor, the principle of indifference, and Huffman codes (Solomonoff, 1964: 5).

2.) Applications Model: Application of methods to specific problems and compare with intuitive evaluations (Solomonoff, 1964: 5).

3.) Goodness Criterion Model: Data suggests one model is as good as another (Solomonoff, 1964: 5)

Occasm's razor is the method of 'the most simple' or 'the most efficient or economical' of several hypothesis. If we apply this using Solomonoff's Model (Chaitin, 1975b: 49), as seen in Example A, we get a theory that is very economical, but is invalid as a generator of the desired series of bits.

Example A:

Original:

Predictions	Number of Spaces
01010101010101010101	20
01010101010000000000	20

If we take each of the number strings from the Solomonoff Model, an using only the first two binary bits in each string, the following would result:

A.) 01

B.) 01

If we used an algorithm based on these two binary bits, the following 'predictive' strings would result:

A.) 01010101010101010101

B.) 01010101010101010101

String [A] is correct.

String [B] is incorrect.

The smallest algorithm did not predict correctly string [B].

A similar case can be made using a Turing machine as can be seen in Example B.

Example B:

Original:

Predictions	Number of Spaces
01010101010101010101	20
01010101010000000000	20

If we take each of the number springs from the Solomonoff Model, an using only the first two binary bits in each string, the following would result:

A.) 01

B.) 01

If we used an algorithm based on these two binary bits, the following 'predictive' strings would result:

A.) 01010101010101010101

B.) 01010101010101010101

String [A] is correct.

String [B] is incorrect.

The smallest algorithm did not predict correctly string [B].

Hence the need for 'types' of information content, not just amounts of information, when coding, and decoding, using a Turing machine.

Using Huffman codes, see Example C, that result in the most 'minimal', an using the reverse of a Huffman code, first obtain a minimal code for a string, and from this obtain the probability of that string (Solomonoff, 1964: 4).

Example C:

Original:

Predictions	Number of Spaces
01010101010101010101	20
01010101010000000000	20

If we take each of the number strings from the Solomonoff Model, an using only the first two binary bits in each string, the following would result:

    A.) 01

    B.) 01

Bradley S. Tice

If we used an algorithm based on these two binary bits, the following 'predictive' strings would result:

    A.) 01010101010101010101

    B.) 01010101010101010101

String [A] is correct.

String [B] is incorrect.

Using the Huffman code results in a correct, i.e accurate, 'minimal' code for string [A], the same cannot be said for string [B]. Even if we use the 'space multiplier' program, the probability of either string [A] or string [B] is not so much a 'probability' as it is a 'reality'. In other words, the minimal code is just a 'short hand' for the longer code, and not an amplifier for stochastic probabilities. If you know the number, sequence, and type of code in advance, there is no probability. There is only code.

The Applications Model is the result of testing the General Intuitive Model's validity. Occam's razor is not the best method of defining information content because the most simple is not necessarily the most accurate when coding. The Turing machine is not a 'crystal ball' and needs information to copy.

Transcription is code.

Accurate transcription is the ideal. The Turing machine must have enough information to duplicate a symbol, or it will not, resulting in an error. The Huffman code is an attempt at 'economy' but is used inversely to predict probability in a code. If one knows the code the probability of reproducing that code is 100%. What is the point to finding the probability of a known code?

The Goodness Criteria Model is based on a very large table of data that can be used by any model with equal probability of results. None of the models proposed by Solomonoff resulted in an accurate series of bits.

In Chaitin's paper "On the Difficulty of Computations" (1970) he states:

In the traditional foundations of the mathematical theory of probability, as expounded by Kolmogorov in his classic [Foundations of the Theory of Probability, 1950], there is no place for the concept of an individual random sequence of 0's and 1's. Yet it is not altogether meaningless to say that the sequence

[A] 11001011111001100101111 0000010

is more random or patternless than the sequences

[B] 1111111111111111111111 111111111

[C] 0101010101010101010101 01010101

for we may describe these last two sequences as thirty 1's or fifteen 10's, but there is no shorter way to specify the first sequence than by just writing it all out (Chaitin, 1970: 6).

Symbolic Space Multiplier:

If we reduce the first sequence of numbers [A] to a more efficient binary bit code by using a space, i.e.. neither a 0 or a 1, as a signal for the computer to multiply the last number before the space, by the coded bit number that follows that space, hence a multiplication program that is formalized by four rules:

1.) The number before the space is the number to be multiplied.

2.) The code bit number following the space is the multiplier.

3.) Two spaces concludes further multiplication procedures and hence returns the computer to the next operation.

4.) The multiplier is designed by a single or multiple character digit code:

Multiplier	Code
4	0
5	1

Sequence [A] can be reduced from the original:

[A] 110010111110011001011110000010

reduces to 8 sets of 4 characters and one set of 3 characters:

[1100] [1011] [1110] [0110] [0101] [1110] [1111]
[0000] [010]

reduced to 17 sets of 2 characters each and 1 set of 1
bit:

[11] [00] [10] [11] [11] [10] [01] [10] [01] [01]
[11] [10] [11] [11] [00] [00] [01] [0]

The most expedient pattern of the sequence to
utilize is the large repetition segments Original:

[A] 1100101111100110010111110000010

Use the Symbolic Space Multiplier program on all
5 sequential character series in the string [A].

[A] 1100101 10011001011110 110

Code Key: Code bit 1=multiplied by 5

This is important because as Chaitin has stated:

We [Kolmogorov] believe that the random or patternless sequences of a given length are those that require the longest programs. We have seen that most of the binary sequences of length k require the longest programs. We have seen that most of the binary sequences of length k require programs of about length k. These then are the random or patternless sequences. Those sequences that can be obtained by putting into a computer a program much shorter than k are the nonrandom sequences, those that possess a pattern or follow a law. The more possible it is to compress a binary sequence into a short program calculation, the less random the sequence (Chaitin, 1970: 6).

With the use of the Symbolic Space Multiplier program string [A] was reduced from the original 30 character number count to a 22 character number count. This qualifies for the [k] length requirement to be less than the original. This satisfies the qualification as a 'nonrandom' or 'patterned' sequence as defined by the equation:

L (M, S) less than or equal to k+1 for all binary sequences S of length k

Key:

M= computer

S= sequence

k= length of sequence

Now while it is true that the ability to compress is a trait of nonrandom series of numbers, the act of compression is not constrained to just nonrandom number series, as has been shown previously, but also the concept of randomness becomes less clear as the formal rule that [k], length of a series, is reducible if nonrandom is not valid, as has been shown in the Symbolic Space Multiplier program.

Of this it can be inferred:

1.) Randomness traits are not clear or well defined.

Bradley S. Tice

2.) Formal rules to random and nonrandom properties

are ambiguous.

Summary

Information theory is the bases of our modern information age. This information model is addressed in context to language, specifically aspects of phonology and redundancies in the English language.

The following are the main points resulting from the analysis in part two of the dissertation in addressing algorithmic information theory:

1.) The ability to compress is a trait of nonrandom series of numbers. The act of compression is not constrained to just nonrandom number series.

2.) It is the quality of information in an algorithm that is important, not the brevity of it, as such information is both quantitative, i.e. necessitates a

certain amount of information to function, as well

as qualitative, i.e. must accurately represent the

information being imparted.

It can be seen that both information theory and

algorithmic information theory are fundamental

aspects of modern technology and that they form an

important field of study in the engineering disciplines.

Appendix A

"Sub-Dividing of a Randon String of Numbers into Subgroups of Uniform Bits that Gives the Impression of Nonrandomness".

Submitted to <u>ACM Transactions on Computational Logic</u> on August 15, 2001

<u>Sub-Dividing of a Random String of Numbers into Subgroups of Uniform Bits that Gives the Impression of Nonrandomness</u>

Bradley S. Tice

P.O. Box 2214 Cupertino, California 95015-2214 USA

Key Words: Randomness, Complexity, Program Size, Algorithms, and Computer Science

Bradley S. Tice

Abstract

The observation that a truly random sequence will exhibit local nonrandomness can be said of the sub-dividing of a random string of numbers into subgroups of uniform bits that gives the impression of nonrandomness. This maybe stated as a 'natural' consequence of reduction to a 'uniform' or patterned sequence of a string of units or bits.

Algorithmic Information Theory (AIT) was invented by A.N. Kolmogorov and developed by R.J. Solomonoff and G. Chaitin. While Solomonoff defines that program-size complexity quantifies Occam's Razor by providing a numerical measure of the degree of simplicity of a scientific theory, it was Kolmogorov and Chaitin that, independently, came up with program-size complexity [1]. Program-size complexity can be stated that if a random string is incompressible, its program-size complexity is as large as possible for bit strings having that length [2].

The observation that a truly random sequence will exhibit local nonrandomness can be said of the subdividing of a random string of numbers into subgroups of uniform bits that gives the impression of nonrandomness. This maybe stated as a 'natural' consequence of reduction to a 'uniform' or patterned sequence of a string of units or bits.

The importance of Chaitin's work is that his theory of complexity-based definition of randomness works for both finite and infinite strings and is equivalent to

Bradley S. Tice

Martin-Lof's (1966) theory for infinite strings [3, 4, & 5].

Kolmogorov's paper (1965) describes the algorithmic approach as "the quantity of information "conveyed by an object" (x) "about an object" (y). It is not an accident that in the probabilistic approach this has led to a generalization to the case of continuous variables, for which the entropy is infinite, but in large number of cases, is finite. The real objects that we study are very (infinitely) complex, but the relationships between two separate objects diminish as the schemes used to describe them becomes simpler" [6].

Some interesting questions arise when reviewing Gregory J. Chaitin's paper "Randomness and Mathematical Proff" (1975b). From Chaitin:

> "...this "incompressibility" is a property of all random numbers; indeed, we can proceed directly to define randomness in terms of incompressibility: A series of number is random if the smallest algorithm capable of specifying it to a computer has about the same number of bits of information as the series itself" [7].

Example (Taken from Chaitin [8]:

Type	Numbers	Reduced
[Predictable]	01010101010101010101	[10 ten times]
[Random]	01101100110111100010	[same amount]

What if we grouped the numbers/bits into a series of 5 groups containing 4 numbers/bits? For ease of discussion the predictable group of numbers will be labelled A.) and the random group numbers B.).

A.) [0101][0101][0101][0101][0101]

B.) [0110][1100][1101][1110][0010]

We can reduce each group of numbers/bits into a specific type of grouping by using an alpha code to represent each type of frequency of numeration.

A.) a=0 followed by a 1, or ax5=01x10, or A.).

B.) a=0110

   b=1100

   c=1101

   d=1110

   e=0010

or a+b+c+d+e=B.).

Reducing both A.) and B.) to a two bit series into an alpha code would produce:

A.) a=0 followed by a 1, or ax10.

B.) a=01

    b=10

    c=11

    d=00

    or [a+b][c+d][c+a][c+b][d+e]

    or a+b+c(d+a+b)+d+e

Notice that although A.) stays the same, i.e. A=ax10, B is reduced from a 20 number/bit series into a group of 5 quad numbered groupings of numbers/bits that is further reduced to a binary code of numbers/bits using 4 alpha symbols to represent the four types of variations of the binary bits, i.e. 00, 11, 01, and 10, that when expressed mathematically combines the three internal groups, all designated by the alpha variable [c] in the head position, that reduces it into an equation of 3 groups from the original 'random' 20 numbers/bits series.

Of note is that if we reduce the 20 number/bit random series [B] into a 4 group alpha series of binary code states, i.e. [00], [11], [01], and [10], the level of randomness or 'uncertainty' is lowered as [c], number 1 followed by number 1 [11], repeats in the primary or head position in the middle three groups in [B].

Knuth (1998) makes the observation that a truly random sequence will exhibit local nonrandomness [9]. So can be said of the sub-dividing of a random string of numbers into subgroups of uniform bits that gives the impression of nonrandomness. This maybe stated as a 'natural' consequence of reduction to a 'uniform' or patterned sequence of a string of units or bits. We can formally propose such a process as:

> A string of numbers that is random in nature will, upon sub-dividing into multiple equal units of measure of the whole, will produce a pattern of bits, that upon inspection, will have a uniform pattern to each sub-group of the divided whole.

Kolmogorov makes a similar observation in his paper (1969) when he states that "It is understood that the description of segments of an infinite sequence of such a nature can be significantly simplified in comparison with the standard description" [10].

In conclusion, the observation that a truly random sequence will exhibit local nonrandomness can be said of the sub-dividing of a random string of numbers into subgroups of uniform bits that gives the impression of nonrandomness. This maybe stated as a 'natural' consequence of reduction to a 'uniform' or patterned sequence of a string of units or bits.

References

[1] G.J. Chaitin, G.J., The Unknowable, Springer-Verlag,. Singapore, 1999.

[2] G.J. Chaitin, G.J., The Unknowable, Springer-Verlag,. Singapore, 1999.

[3] G.J. Chaitin, G.J., The Unknowable, Springer-Verlag,. Singapore, 1999.

[4] G.J. Chaitin, "A theory of program size formally identical to information theory", **Journal of the Association for Computing Machinery, 22** (1975a), 329-340.

[5] P. Martin-Lof, "The definition of random sequences", **Information and Control, 9** (1966), 602-619.

[6] A.N. Kolmogorov, "Three approaches to the quantitative definition of information", **Problems of Information Transmission, 1** (1965), 1-7.

[7] G.J. Chaitin, "Randomness and Mathematical Proof", **Scientific American, 232** (1975b) 47- 52.

[8] G.J. Chaitin, "Randomness and Mathematical Proof", **Scientific American, 232** (1975b) 47- 52.

[9] D. Knuth, The Art of Computer Programming, Addison-Wesley, Menlo Park, California, 1998.

[10] A.N. Kolmogorov, "On the logical foundations of information theory and probability theory", **Problems of Information Transmission, 5** (1969), 1-4.

Appendix B

"Indefinite Loops and the Undecidability of the Halting Problem Using Algorithmic Information Theory". Submitted to the <u>Nordic Journal of Computing (NJC)</u> on December 12, 2000.

Infinite Loops and the Undecidability of the Halting Problem Using Algorithmic Information Theory

Bradley S. Tice

AHD

P.O. Box 2214

Cupertino, CA 95015-2214 USA

Bradley S. Tice

Key Words: Algorithmic Information Theory, Undecidability, Numerical Analysis, Computer Science, and Recursion Theory

Abstract

The use of algorithmic information theory to address the undecidability of the halting problem that uses Chaitin's EVAL program as a constraint to an infinite loop must be considered an indefinite process to this system.

Chaitin has defined his algorithmic information theory to address the undecidability of the halting problem. The Algorithmic Information Theory Theorem is as follows:

The base-two representation of the probability that U halts is a random (i.e. maximally complex) infinite string [1].

Chaitin does this by using John McCarthy's LISP program (1960) but develops beyond McCarthy's program in three fundamental ways [2, 3]:

1) Simplify LISP by only allowing atoms to be one character long.

2) EVAL must not lose control by going into an infinite loop. EVAL must execute garbage for a limited amount of time and always results in an error message or a valid value or expression.

3) Stipulate a 'permissive' LISP semantics with the property that the only way a syntactically valid LISP expression can fail to have a value if it loops forever.

Bradley S. Tice

Notice that EVAL, as set up by Chaitin, allows for a limited amount of calculation of an 'undefined' function without going into an infinite loop. In other words, Chaitin has allowed for a limited amount of calculation to an 'undefined' function without it becoming obsessive, i.e. approaching an infinity, although it is unknown what, and or if, such a calculation will result in defining an output for this function. Also note that in the third requirement for Chaitin's LISP program, the only way a syntactically valid LISP expression can fail is if it loops forever.

The criteria for an infinite loop is not that it loops forever, i.e. an infinity, but that it loops, i.e. calculates an algorithmic recursive function indefinitely, for a specific number of times, finite, and then decides that it is an 'undefined' function, i.e. that it is not a function because time and process duration is exceeded for the calculation on whether it halts or not.

Again, another way of bypassing the halting problem and assigning a false value, i.e. an untrue value to a mathematical, or computational, value in that an undefined set or program is just that, undefined or indefinite. It is not known if it is infinite. Assigning an infinite loop to an indefinite loop is inaccurate and undermines the vary bases of scientific and logical thought.

References

[1] Chaitin, G.J. 1975 "A theory of program size formally identical to information theory". **Journal of the Association for Computing Machinery**, 1975, Volume 22, pp. 329- 340.

[2] McCarthy, J. 1960 "Recursive functions of symbolic expressions and their computation by machine, Part 1". **Communications of the ACM**, April 1960, Volume 3, Number 4, pp. 184-195.

[3] Chaitin, G.J. 1987 Algorithmic Information Theory, Cambridge: Cambridge University Press.

References

Abramson, N. 1963 <u>Information Theory and Coding</u>.

New York: McGraw Hill.

Anderson, S.R. 1985 <u>Phonology in the Twentieth Century.</u>

Chicago: The University of Chicago Press.

Ash, R. 1965 <u>Information Theory.</u>

New York: Interscience Publishers.

Beltrami, E. 1999 <u>What is Random?</u>

New York: Copernicus.

Bennett, D.J. 1998 <u>Randomness</u>.

Cambridge: Harvard University.

Brown, P.F. and Della Pietra, S.A. 1992 "An estimate of an upper

bound for the entropy of english". **Computational**

Linguistics, March 1992, Volume 18, Number 1, pages 31-40.

Birdwhistell, R.L. 1970 <u>Kinesics and Context.</u>

Philadelphia: University of Pennsylvania Press.

Chaitin, G.J. 1966 "On the length of programs for computing finite binary sequences". **Journal of the Association for Computing Machinery**, Volume 13, No. 4, October 1966, pp.547-569.

Chaitin, G.J. 1970 "On the difficulty of computations".

Information Theory, January 1970, Volume 16, Number 1, pp. 5-9.

Chaitin, G.J. 1975a "A theory of program size formally identical to information theory". **Journal of the Association for Computing Machinery**, 1975, Volume 22, pp. 329-340.

Chaitin, G.J. 1975b "Randomness and Mathematical Proof"

Scientific American, May 1975, Volume 232, Number 5,

pp. 47-52.

Chaitin, G.J. 1999 The Unknowable.

Singapore: Springer-Verlag.

Cherry, E.C., Halle, M., and Jakobson, R. 1953 "Toward the

logical description of languages in their phonemic aspect".

Language, Volume 29, 1953, pages 34-46.

Church, A. 1940 "On the concept of a random sequence".

Bulletin of the American Mathematical Society, December

1940, Volume 46, Number 12, Part 2, pp. 130-135.

Fano, R.M. 1961 Transmission of Information.

New York: The MIT Press and John Wiley & Sons.

Gilbert, E.N. 1966 "Information theory after 18 years".

Bradley S. Tice

   Science, April 15, 1966, pages 320-326.

Kahn, D. 1967/1996 <u>The Code Breakers: The Story of Secret
   Writing.</u> New York: Scribner.

Knuth, D. 1998 <u>The Art of Computer Programming.</u>
   Menlo Park, California: Addison-Wesley.

Kolmogorov, A.N. 1965 "Three approaches to the quantitative
   definition of information". **Problems of Information
   Transmission**, January-March 1965, Volume 1, Number 1,
   pp. 1-7.

Kolmogorov, A.N. 1969 "On the logical foundations of
   information theory and probability theory". **Problems of
   Information Transmission**, July-September 1969, Volume
   5, Number 3, pp. 1-4.

Manning, C.D. and Schutze, H. 1999 <u>Foundations of Statistical
   Natural Language Processing</u>. Cambridge: The MIT Press.

Martin-Lof, P. (1966) "The definition of random sequences".

Information and Control, 1966, Volume 9. pages 602-619.

Shannon, C.E. 1950/1993 "Prediction and entropy of printed

english". In Claude Elwood Shannon Collected Papers edited

by Sloane, N.J.A. and Wyner, A.D. New York: The Institute

of Electrical and Electronics Engineers, Inc. (IEEE) Press.

Shannon, C.E. and Weaver, W. (1949) The Mathematical Theory

of Communication. Urbana: University of Illinois Press

Siegfried, T. 2000 The Bit and the Pendulum.

New York: John Wiley & Sons, Inc.

Solomonoff, R.J. 1964 "A formal theory of inductive inference".

Information Control, Volume 7, March 1964, pp. 1-22, and

June 1964, pp. 224-254.

Bradley S. Tice

Tice, B. 1997 "Language Learning Loop: A Pronunciation System for Japanese ESL" in <u>TESOL Matters</u> Volume 7, Number 2, April/May 1997, page 12.

Tice, B. 1998 "Feedback Systems for Nontraditional Medicines: A Case for the Signal Flow Diagram" in the <u>Journal of Pharmaceutical Sciences,</u> November 1998, Volume 87, Number 11, pp. 1282-1285.

von Neumann, J. and Morgenstern, O. 1944 <u>Theory of Games and Economic Behavior</u>. New York: John Wiley & Sons, Inc.

About the Author

Dr. Tice is Director and Institute Professor of Chemistry at Advanced Human Design in Cupertino, California U.S.A. Dr. Tice is currently doing research on organic semiconductors.

Postscript

Entered this dissertation into the 2001 ACM, Association for Computing Machinery, Doctoral Dissertation Competition.

* Association for Computing Machinery (ACM)
1515 Broadway, New York, New York 10036-5701